THE
SERVICEBERRY

ABUNDANCE and RECIPROCITY
in the NATURAL WORLD

ROBIN WALL KIMMERER

WITH ILLUSTRATIONS BY JOHN BURGOYNE

SCRIBNER

New York   Toronto   London   Sydney   New Dehli

Scribner
An Imprint of Simon & Schuster, LLC
1230 Avenue of the Americas
New York, NY 10020

First Scribner hardcover edition November 2024

SCRIBNER and design are trademarks of Simon & Schuster, LLC

Simon & Schuster: Celebrating 100 Years of Publishing in 2024

For information about special discounts for bulk purchases,
please contact Simon & Schuster Special Sales at 1-866-506-1949
or business@simonandschuster.com.

The Simon & Schuster Speakers Bureau can bring authors to
your live event. For more information or to book an event,
contact the Simon & Schuster Speakers Bureau at 1-866-248-3049
or visit our website at www.simonspeakers.com.

Interior design by Kyle Kabel

Manufactured in the United States of America

5   7   9   10   8   6

Library of Congress Control Number: 2024943789

ISBN 978-1-6680-7224-0
ISBN 978-1-6680-7225-7 (ebook)

"The Serviceberry" was originally published,
in an abbreviated form, in *Emergence Magazine*.

To my good neighbors,
Paulie and Ed Drexler.

All
Flourishing
Is
Mutual

THE COOL BREATH of evening slips off the wooded hills, displacing the heat of the day, and with it come the birds, as eager for the cool as I am. They arrive in a flock of calls that sound like laughter, and I have to laugh back with the same delight. They are all around me, Cedar Waxwings and Catbirds and a flash of Bluebird iridescence. I have never felt such a kinship to my namesake, Robin, as in this moment when we are both stuffing our mouths with berries and chortling with happiness. The bushes are laden with fat clusters of red, blue, and wine purple in every stage of ripeness—so many, you can pick them by the handful. I'm glad I have a pail, and it's getting pretty heavy. The birds carry their berries in the buckets of their bellies and wonder if they will be able to fly with so much cargo.

This abundance of berries feels like a pure gift from the land. I have not earned, paid for, nor labored for them. There is no mathematics of worthiness that reckons I deserve them in any way. And yet here they are—along with the sun and the air and the birds and the rain, gathering in towers of cumulonimbi, a distant storm building. You could call them natural resources or ecosystem services, but the Robins and I know them as gifts. We both sing gratitude with our mouths full.

Part of my delight comes from their unexpected presence. I never imagined that I could pick them here. The local native Serviceberries, *Amelanchier arborea,* have small, hard fruits, which tend toward dryness, and only once in a while is there a tree with sweet offerings. The bounty in my bucket today is a western species—*A. alnifolia,* known as Saskatoons—planted by my farmer neighbors, Paulie and Ed. This is their first bearing year, and they produce berries with an enthusiasm that matches my own.

Saskatoon, Juneberry, Shadbush, Shadblow, Sugarplum, Sarvis, Serviceberry—these are among the many names for *Amelanchier*. Ethnobotanists know that the more names a plant has, the greater its cultural importance. The tree is beloved for its fruits, for medicinal use, and for the early froth of flowers that whiten woodland edges at the first hint of spring. Serviceberry is known as a calendar plant, so faithful is it to seasonal weather patterns. Its bloom is a sign that the ground has thawed. In this folklore, this was the time that mountain roads became passable for circuit preachers, who arrived to conduct church services. It is also a reliable indicator to fisherfolk that the shad are running upstream—or at least it was back in the day when rivers were clear and free enough to support the spawning of shad.

Calendar plants like Serviceberry are important for synchronizing the seasonal rounds of traditional

Indigenous People, who move in an annual cycle through their homelands to where the foods are ready. Instead of changing the land to suit their convenience, they changed themselves. Eating with the seasons is a way of honoring abundance, by going to meet it when and where it arrives. A world of produce warehouses and grocery stores enables the practice of having what you want when you want it. We force the food to come to us, at considerable financial and ecological costs, rather than following the practice of taking what has been given to us, each in its own time. These Serviceberries were not coerced, and their carbon footprint is nothing. Maybe that's why they taste so good—they come only this time of year—these ephemeral sips of summer, without the aftertaste of harm.

The name "Serviceberry" comes not from its "service" but from a very old version of its Rose Family name, "Sorbus," which became "sarvis" and hence "service." While the name did not derive from its benefits, the plant does provide myriad goods and services—not only to humans but to many other citizens. It supports biodiversity. Shadbush is a preferred browse of Deer and Moose, a vital source of early pollen for newly emerging insects, and host to a suite of butterfly larvae—like Tiger Swallowtails, Viceroys, Admirals, and Hairstreaks—and berry-feasting birds who rely on those calories in breeding season.

Human people, too, rely on those calories, especially in traditional Indigenous food practices. Serviceberries were a critical ingredient in the making of pemmican. The dried berries, along with dried venison or bison, were pounded to a fine powder, bound with rendered fat, and solidified into the original energy bars. This highly concentrated, preserved food provided full nutritional sustenance through seasons of hunger, was easily transportable, and could be cached or carried. Pemmican became part of the traditional trade economy, a sophisticated local and transcontinental network that distributed vital materials across ecosystems and cultures. Surplus Serviceberry calories could be exchanged for other goods not locally available.

Serviceberries are part of Indigenous foodways wherever they grow. I am a member of the Potawatomi Nation, which is one of the Anishinaabe peoples of the Great Lakes region. I've had the privilege of eating syrupy purple compotes of Serviceberries at traditional feasts, which have primed my taste buds and my memories of this ancestral food.

In Potawatomi, it is called *Bozakmin,* which is a superlative: the best of the berries. I feel one on my tongue, and agree with my ancestors on the rightness of that name. Imagine a fruit that tastes like a Blueberry crossed with the satisfying heft of an Apple, a touch of rosewater, and a minuscule crunch of almond-flavored seeds. They taste like nothing a grocery store has to offer: wild, complex with a flavor that your body recognizes as the real food it's been waiting for. I can almost feel my mitochondria doing a happy dance when I eat them.

For me, the most important part of the word *Bozakmin* is "min," the root for "berry." It appears in our Potawatomi words for Blueberry (*Minaan*), Strawberry (*Odemin*), Raspberry (*Mskadiismin*), even Apple (*Mishiimin*), Maize (*Mandamin*), and Wild Rice (*Manomin*). That word is a revelation, because it is also the root word for "gift." In naming the plants who shower us with goodness, we recognize that these are gifts from our plant relatives, manifestations of their generosity, care, and creativity. James Vukelich, an Anishinaabe linguist, teaches that these plant gifts are "a manifestation of unconditional love that plants have for people." Plants offer whatever they have, to whoever needs it, "saint and sinner, alike," he writes.

I can't help but gaze at them, these shiny gems, cupped in my hand—and breathe out my thanks. In the presence of such gifts, gratitude is the intuitive first response. This gratitude flows toward our plant elders and radiates to the rain, to the sunshine, to the improbability of bushes spangled with morsels of sweetness in a world that can be bitter.

In the Anishinaabe worldview, it's not just fruits that are understood as gifts, rather all of the sustenance that the land provides, from fish to firewood. Everything that makes our lives possible—the splints for baskets, roots for medicines, the trees whose bodies make our homes, and the pages of our books—is provided by the lives of more-than-human beings. This is always true whether it's harvested directly from the forest or whether it's mediated by commerce and harvested from the shelves of a store—it all comes from the Earth. When we speak of these not as things or natural resources or commodities, but as gifts, our whole relationship to the natural world changes.

In a traditional Anishinaabe economy, the land is the source of all goods and services, which are distributed in a kind of gift exchange: one life is given in support of another. The focus is on supporting the good of the people, not only an individual. Receiving a gift from the land is coupled to attached responsibilities of sharing, respect, reciprocity, and gratitude—of which you will be reminded.

This kind of gratitude is so much more than a polite "thank you." Not an automatic ritual of "manners," but a recognition of indebtedness that can stop you in your tracks—it brings you the realization that your life is nurtured from the body of Mother Earth. With my fingers sticky with berry juice, I'm reminded that my life is contingent upon the lives of others, without whom, I simply would not exist. Water is life, food is life, soil is life—and they become our lives through the paired miracles of photosynthesis and respiration. All that we need to live flows through the land. It is not an empty metaphor that we call her Mother Earth. Food in our mouths is the thread that connects us in a relationship simultaneously spiritual and physical, as our bodies get fed and our spirits nourished by a sense of belonging, which is the most vital of foods.

I have no claim to these berries, and yet here they are in my bucket, a gift.

This pail of Juneberries represents hundreds of gift exchanges that led up to my blue-stained fingers: the Maples who gave their leaves to the soil, the countless invertebrates and microbes who exchanged nutrients and energy to build the humus in which a Serviceberry seed could take root, the Cedar Waxwing who dropped the seed, the sun, the rain, the early spring flies who pollinated the flowers, the farmer who wielded the shovel to tenderly settle the seedlings. They are all parts of the gift exchange by which everyone gets what they need.

Many Indigenous Peoples, including my Anishinaabe relatives and my Haudenosaunee neighbors, inherit what is known as "a culture of gratitude," where lifeways are organized around recognition and responsibility for earthly gifts, both ceremonial and pragmatic. Our oldest teaching stories remind us that failure to show gratitude dishonors the gift and brings serious consequences. If you dishonor the Beavers by taking too many they will leave. If you waste the Corn, you'll go hungry.

Enumerating the gifts you've received creates a sense of abundance, the knowing that you already have what you need. Recognizing "enoughness" is a radical act in an economy that is always urging us to consume more. Data tell the story that there are "enough" food calories on the planet for all 8 billion of us to be nourished. And yet people are starving. Imagine the outcome if we each took only enough, rather than far more than our share. The wealth and security we seem to crave could be met by sharing what we have. Ecopsychologists have shown that the practice of gratitude puts brakes on hyperconsumption. The relationships nurtured by gift thinking diminish our sense of scarcity and want. In that climate of sufficiency, our hunger for more abates and we take only what we need, in respect for the generosity of the giver. Climate catastrophe and biodiversity loss are the consequences of unrestrained taking by humans. Might cultivation of gratitude be part of the solution?

Paulie is surprised, too, that I know these berries since they are new to most people around here. As a forager, I'm accustomed to following the voices of Cedar Waxwings, only to find a disappointing handful that they have left me. Never have I seen so many as here, planted by my friend. She's excited to learn that they are an important cultural food for us, and I can see how my delight levitates her, too.

If our first response to the receipt of gifts is gratitude, then our second is reciprocity: to give a gift in return. What could I give these plants in return for their generosity? I could return the gift with a direct response, like weeding or bringing water or offering a song of thanks that sends appreciation out on the wind. I could make habitat for the solitary bees that fertilized those fruits. Or maybe I could take indirect action, like donating to my local land trust so that more habitat for the gift givers will be saved, speaking at a public hearing on land use, or making art that invites others into the web of reciprocity. I could reduce my carbon footprint, vote on the side of healthy land, advocate for farmland preservation, change my diet, hang my laundry in the sunshine. We live in a time when every choice matters.

Gratitude and reciprocity are the currency of a gift economy, and they have the remarkable property of multiplying with every exchange, their energy concentrating as they pass from hand to hand, a truly renewable resource.

Can we imagine a human economy with a currency which emulates the flow from Mother Earth? A currency of gifts?

When I speak about reciprocity as a relationship, let me be clear. I don't mean a bilateral exchange in which an obligation is incurred, and can then be discharged with a reciprocal "payment." I mean keeping the gift in motion in a way that is open and diffuse, so that the gift does not accumulate and stagnate, but keeps moving, like the gift of berries through an ecosystem. We ecologists think about the currency of ecosystems in terms of biogeochemistry—the cycling of life's materials, between the living and the not.

The berries make a satisfying "plunk" in my bucket. It's worth thinking about what they are made of as the pail gets heavier. Those Service-berries contain both elemental materials, like carbon and nitrogen, and the energy stored in the sweet sugars. To understand this natural economy and apply it to our own, we have to remember that materials and energy move differently through an ecosystem.

Materials like carbon and nitrogen and phosphorus—the essential elements of life—cycle through an ecosystem, endlessly changing form as they are passed along. Let's follow the carbon in that Saskatoon. The leaves of the tree drew carbon dioxide from the atmosphere, which they made into sugar via the brilliant mechanism of photosynthesis. The gift of the atmosphere now resides in the berry. When the Cedar Waxwing gobbles it up, some of that carbon becomes the feathers that paint a yellow band on its tail, which flashes in the afternoon light. When that feather falls to the ground it becomes food for beetles, who become food for a Vole whose death feeds the soil who feeds the Serviceberry seedling just germinating at the edge of the woods. Materials move through ecosystems in a circular economy and are constantly transformed. Abundance is created by recycling, by reciprocity.

This recycling proceeds at different paces. Sometimes it is as quick as minutes, like a molecule of phosphorus dancing between water and a spinning cell of virid green alga. That alga takes up the phosphorus into its body, which is eaten by zooplankton a few minutes later, and they excrete the mineral back into the water, where another alga is happy to have it. Other cycles proceed more slowly. Sometimes the minerals get squirreled away in long-term storage, like nitrogen immobilized in a tree trunk for three hundred years, but it always comes back into circulation. The juice that bursts from these berries was rain just last week and is already on its way back to the clouds. These processes are the models for principles of a circular economy, in which there is no such thing as waste, only starting materials. Abundance is fueled by constantly circulating materials, not wasting them.

Energy, however, is a very different story. While chemical materials can cycle in an ecosystem, energy flows in one inevitable direction. Even though it can be temporarily stored, it always moves on, thanks to the laws of thermodynamics. The Sun's energy, stored in the chemical bonds of a Serviceberry, will fuel the trilling voices of the Waxwings, but will eventually be dissipated as heat emanating from the warm feathered body. Energy cannot ever be completely recycled; it gets used up in the thermodynamic inefficiency of energy transfer among beings. Therefore, energy must be constantly replenished to fuel the flow. I have solar panels on my roof and my life is as thoroughly solar-powered as they are. Is it any wonder that the Sun has always been revered as the source of life?

I suppose in an industrial economy "production" is the source of the flow, rooted in human labor and the conversion of earthly gifts to commodities. But so often that production is at the cost of great destruction. When an economic system actively destroys what we love, isn't it time for a different system?

Some powerful feminist thinkers call us to remember that gift giving is among the most primal of human relationships. Each of us begins our life as the recipient in what Genevieve Vaughan has called a "maternal gift economy," the flow of "goods and services" from mother to newborn. When the mother nurses her child, the boundary of the individual self becomes permeable and the common good is the only one that matters. The maternal gift economy is a biological imperative. There is no meritocracy or earning of sustenance.

Mothers do not sell their milk to their babies, it is pure gift, so that life can continue. The currency of this economy is the flow of gratitude, the flow of love, literally in support of life.

By analogy, can the sustenance from the breast of Mother Earth be understood as a maternal gift economy? These feminist thinkers argue that giving and taking in this sense are a fundamental way of caring for each other, without the intervention of states or markets. Scholars like Miki Kashtan are exploring how the philosophy and practice of a maternal gift economy might move social organization toward justice and sustainability.

If the Sun is the source of flow in the economy of nature, what is the "Sun" of a human gift economy, the source that constantly replenishes the flow of gifts? Maybe it is love.

IN A SERVICEBERRY ECONOMY, I accept the gift from the tree and then spread that gift around, with a dish of berries to my neighbor, who makes a pie to share with his friend, who feels so wealthy in food and friendship that he volunteers at the food pantry. You know how it goes.

In contrast, if I were to buy a basket of berries in a market economy, the relationship ends with the exchange of money. Once I hand over my credit card, I have no further exchange with the clerk or the store. We're done. I own these berries now and can do with them as I like. The clerk, the corporation, and I—the customer—have a strictly material transaction. There is no making of community, only a trading of commodities. Think of how strange—but wonderful—it would feel if you met the clerk on the street and they asked for your recipe for Serviceberry pie. That would be out-of-bounds. But if those berries were a gift you'd probably still be chatting.

To name the world as gift is to feel your membership in the web of reciprocity. It makes you happy—and it makes you accountable. Conceiving of something as a gift changes your relationship to it in a profound way, even though the physical makeup of the "thing" has not changed. A woolly knit hat that you purchase at the store will keep you warm regardless of its origin, but if it was hand-knit by your favorite auntie, then you are in relationship to that "thing" in a very different way: you are responsible for it, and your gratitude has motive force in the world. You're likely to take much better care of the gift hat than of the commodity hat, because the gift hat is knit of relationships. This is the power of gift thinking. I imagine if we acknowledged that everything we consume is the gift of Mother Earth, we would take better care of what we are given.

I was once giving a lecture in a College of Natural Resources at a big, prestigious university. I took the opportunity to interrogate that name, because, after all, "natural resources" does mean raw materials to be converted into something we value. As it happens, that college was in the process of changing its name, given its extractive implications. So I offered a suggestion: "Why don't you change your name to the Department of Earthly Gifts?" You should have seen the beatific smiles that went around the room. "Ohhh, yes," people said, with palpable yearning, "we want to work for the Department of Earthly Gifts." But, of course, they chose something else. "It's beautiful idea," a colleague told me later, "but it would kind of put the kibosh on grinding up trees."

Mistreating a gift has emotional and ethical gravity as well as ecological resonance. For example, I'm thinking of a spring I know where the water is icy cold and gushes out of the ground. It almost makes you dizzy with its cold vitality. I drink from my cupped hands, splash my face, and fill my canteen for later. Isn't this how water was meant to be—free and pure? How long has it been since you drank wild water? It feels like a gift to me. The life of that water became my life—and my joy in its presence. Gift thinking means that in gratitude for the drink, I'll clean the leaves from the bottom of the pool and take care not to muddy the edges. I care for the gift so it can keep on giving.

But if I mistreated that spring, by peeing in it or by damming up water that belongs to itself and selling it off, the consequences would be an emotional breach as well as the ruination of the water quality. I'd feel as dirty as the water. It's a gut punch to even imagine someone claiming to own it. But that moral sense does not constrain an economy that views water as a commodity, property to be bought and sold. It seems absurd to me that someone could own water, a free gift that falls like the proverbial manna from heaven. Could you sell manna without spiritual jeopardy? I don't think so.

How we think ripples out to how we behave. If we view these berries or that spring as an object, as property, it can be exploited as a commodity in a market economy. When something moves from the status of gift to the status of commodity, we can become detached from mutual responsibility. We know the consequences of that detachment.

Why then have we permitted the dominance of economic systems that commoditize everything? That create scarcity instead of abundance, that promote accumulation rather than sharing? We've surrendered our values to an economic system that actively harms what we love. Our metrics of economic value like GDP count only monetary value in the marketplace, of that which can be bought and sold. There is no room in these equations for the economic value of clean air and carbon sequestration and the ineffable riches of a forest filled with birdsong. Where is the value of a butterfly whose species has prospered for millennia and lives nowhere else on the planet? There is no formula complex enough to hold the birthplace of stories. It pains me to know that an old-growth forest is "worth" far more as lumber than as the lungs of the Earth. And yet I am harnessed to this economy, in ways large and small, yoked to pervasive extraction. I'm wondering how we fix that. And I'm not alone.

BECAUSE I'M A BOTANIST, my knowledge of economics and finance is about the size of the frilly little cup at the tip of a Juneberry that was once part of the flower. It's called the "calyx," in case you were craving a delicious new word, the way some people crave money.

My fluency in the lexicon of berries does not easily extend to economics, so I wanted to revisit the conventional meaning of economics to compare it with my understanding of the gift economy of nature. What is economics for anyway? The answer to that question depends a lot on who you ask. On their website, the American Economic Association says, "It's the study of scarcity, the study of how people use resources and respond to incentives."

My son-in-law Dave teaches high school economics, and the first principle his students learn is that economics is about decision making in the face of scarcity. Anything and everything in a market is implicitly defined as scarce. With scarcity as the main principle, the mindset that follows is based on commodification of goods and services.

I'm way past high school, but I'm not sure I grasp that thinking, so I fill a bowl with fresh Serviceberries for my friend and colleague Dr. Valerie Luzadis. She is an appreciator of earthly gifts and a professor and past president of the U.S. Society for Ecological Economics. Ecological economics is a growing field that integrates Earth's natural systems and human values and ethics into conventional economic theory. Valerie prefers to define economics as "how we organize ourselves to sustain life and enhance its quality. It's a way of considering how we provide for ourselves." I like that better.

The words "ecology" and "economy" come from the same root, the Greek *oikos,* meaning "home" or "household": i.e., the systems of relationship, the goods and services that keep us alive. The system of market economies that we're given as a default is hardly the only model out there. Anthropologists have observed and shared multiple cultural frameworks colored by very different worldviews on "how we provide for ourselves."

As the berries plunked into my bucket, I was thinking about what I'd do with them all. I'd drop some off for friends and neighbors, and I'd certainly fill the freezer for Juneberry muffins in February. This "problem" of deciding what to do with abundance reminds me of a report, shared by Lewis Hyde in his essential book *The Gift*, that linguist Daniel Everett wrote as he was learning from a hunter-gatherer community in the Brazilian rainforest.

He observes that a hunter had brought home a sizable kill, far too much to be eaten by his family. The researcher asked how he would store the excess. Smoking and drying technologies were well known; storing was possible. The hunter was puzzled by the question—store the meat? Why would he do that? Instead, he sent out an invitation to a feast, and soon the neighboring families were gathered around his fire, until every last morsel was consumed. This seemed like maladaptive behavior to the anthropologist, who asked again: given the uncertainty of meat in the forest, why didn't the hunter store the meat for himself, which is what the economic system of his home culture would predict.

"Store my meat? I store my meat in the belly of my brother," replied the hunter.

I feel a great debt to this unnamed teacher for these words. There beats the heart of gift economies, an antecedent alternative to market economies, another way of "organizing ourselves to sustain life." In a gift economy, wealth is understood as having enough to share, and the practice for dealing with abundance is to give it away. In fact, status is determined not by how much one accumulates, but by how much one gives away. The currency in a gift economy is relationship, which is expressed as gratitude, as interdependence and the ongoing cycles of reciprocity. A gift economy nurtures the community bonds that enhance mutual well-being; the economic unit is "we" rather than "I," as all flourishing is mutual.

Anthropologists characterize gift economies as systems of exchange in which goods and services circulate without explicit expectations of direct compensation. Scientist and philosopher Marshall Sahlins names generalized reciprocity as the heart of a gift economy, which functions most effectively in small, close-knit communities. Those who have give to those who don't so that everyone in the system has what they need. It is not regulated from above but derives from a collective sense of equity in "enoughness" and accountability in distributing the gifts of the Earth.

In his book *Sacred Economics,* Charles Eisenstein states: "Gifts cement the mystical realization of participation in something greater than oneself which, yet, is not separate from oneself. The axioms of rational self-interest change because the self has expanded to include something of the other." If the community is flourishing, then all within it will partake of the same abundance—or shortage—that Nature provides.

In a gift economy, the currency in circulation is gratitude and connection rather than goods or money. A gift economy includes a system of social and moral agreements for indirect reciprocity, rather than a direct exchange So, the hunter who shared the feast with you today could well anticipate that you would share from a full fishnet or offer your labor in repairing a boat in the future. The prosperity of the community grows from the flow of relationships, not the accumulation of goods.

When the natural world is understood as a gift instead of private property, there are ethical constraints on the accumulation of abundance that is not yours to own. Gifts are not meant to be hoarded, and thus made scarce for others, but given away, which generates sufficiency for all.

Gift economies, from the informal to highly ritualized protocols, are known from traditional Indigenous communities around the world. Gatherings of our Potawatomi people often include a "giveaway," a ceremony meant to reinforce relationships through gift giving. In the western world, someone celebrating a life event might expect to receive gifts, but in our way the equation is reversed. The ones who have been blessed with good fortune share that blessing by giving away.

Well-known examples of gift economies include the potlatches of Pacific Northwest peoples, in which gifts circulate in the group, solidifying bonds and redistributing wealth. Traditional potlatches are gift-giving celebrations, in which possessions are given away with lavish generosity to mark meaningful life events. The ceremonial feasts display the wealth of the givers, enhance their prestige, and affirm connections with a web of relations. The gifts received are likely to be given away at the next ceremony, keeping wealth in motion and cementing mutual bonds. This ritualized redistribution of wealth was banned by colonial governments, under the influence of missionaries in the 1800s. Potlatches were seen as contrary to "the civilized values of accumulation" and undermined the notions of individual property and advancement essential to assimilation to the colonial agenda.

These traditional values of relationship and reciprocity continue to resonate in contemporary Indigenous economics, as Dr. Ronald Trosper, a Salish-Kootenai economist has documented in his book *Indigenous Economics: Sustaining Peoples and Their Lands*. Making good relationships with the human and more-than-human world is the primary currency of well-being. These relational values shape current agreements regarding a diversity of tribal economic needs from timber to salmon. The questions of land as moral responsibility and land as commodity are sometimes contested in the headlines.

Trosper tells the story of how making relationships led to the historic intertribal agreements with the U.S. government to protect the cultural landscape of the Bears Ears as the first tribally focused national monument. Five different tribes nurtured relationships with the federal government to forever protect an earthly gift to be held in common. This was a transformative step toward healing a long history of colonial taking. That hopeful model of Indigenous economics was abruptly curtailed when Donald Trump reversed the decision and instead conveyed rights to those sacred lands to a private uranium-mining company. It took an election to reverse it.

The economic contest between colonial and Indigenous currencies did not end with the Buffalo.

These two economic worldviews, of prosperity gained through individual accumulation and prosperity gained through sharing of the commons, underpin the history of colonization in this country. The whole enterprise of dispossession and assimilation of the original peoples was designed to eradicate the notion of land as a source of belonging and to replace it with the idea that land is nothing more than a source of belongings. This required a narrowing of the definition of well-being, from common wealth to individual wealth, from abundance to scarcity.

You don't have to participate in a potlatch to experience the gift economy; when you open your awareness and give them a name, you can see gift economies all around you.

W HEN AUGUST ROLLS AROUND, my neighbor Sandy down the road in an old farmhouse puts out a card table under her front maples. The canning jars with stalks of bright gladiolas caught my eye. There is also a stack of zucchini, a basket of new red potatoes, and a sign that says "free." A person can only use so many glads, and so she shares the gladness with whoever might stop. Zucchinis are another story.

Around here, when late summer heat might bring a new squash every day, finding homes for surplus zukes is no joke. They will grow from cucumber size to baseball bats in a matter of days. People have been known to put them in each other's mailboxes or surreptitiously place them on the front seat of a parked car. I'm not sure that this could be framed as a gift—more like a stealthy game of self-preservation. But not everyone has a garden and a pestilence of squash. Sandy gets the pleasure of seeing a car stop, maybe on the way home from work, to accept the gift of fresh veggies for dinner and a bouquet for the table. The currency of exchange is the secret smiles on both faces.

The practice of the front yard giveaway seems to be contagious on our road. One day, an old travel trailer appeared parked at the edge of a newly shorn hayfield. No electricity or water. A week later, a rough table was set up outside, boards between two sawhorses. Arrayed there was an ornate spice rack, a small heap of Army camouflage items, a canvas-wrapped canteen, a knapsack, and a mess kit. And a sign that said "free."

Is this an economy? I think it is—a system of redistribution of wealth based on abundance and the pleasure of sharing. Someone says: I have more than I need, so I offer it to you. I don't think it's a coincidence that these small acts all happen on a few miles of one country road. Giving begets giving and the gift stays in motion. And there are a lot more roads.

In times of crisis the gift economy surges up through the rubble of an earthquake or the wreckage of a hurricane. Rebecca Solnit, in her stunning book *A Paradise Built in Hell,* describes how gift economies seem to arise spontaneously in times of disaster. When human survival is threatened, compassionate acts overrule market economies. People give freely to one another, and bonds of ownership disappear when everyone pools their resources of food and labor and blankets in solidarity. When systems of governance and market economies of debt are disrupted, networks of mutual aid arise. Heroic tales are told of giving away truckloads of bread pulled from bakery shelves. Just hours ago, that bread was private property that would be sold at a profit and defended from thieves, but in the moment of trouble, it becomes a gift. And the giver of hot soup from a shared camp stove on a street corner earned the prestige and honor of a potlatch host. We know how to do this—and what's more, we crave doing it, feeling more alive with every gift exchange.

The challenge is to cultivate our inherent capacity for gift economies without the catalyst of catastrophe. We have to believe in our neighbors, that our shared interests supersede the impulses of selfishness. There is a tragedy in believing the proffered narrative of our system, which turns us against each other in a zero-sum game.

I was surprised to learn that one of the assumptions of modern economic theory is that we will each behave in conformity to Adam Smith's "Rational Economic Man," characterized as a greedy, isolated individual acting purely in self-interest to maximize return on investment. The system is designed to support this hypothetical caricature and seems therefore to produce them. But each of us knows that exceptions to this behavior far outnumber the predicted stereotype. Can we imagine a system which nurtures a different economic identity and reclaim ourselves as neighbors, with shared investment in mutual well-being? As it happens, abundant empirical evidence attests to the fact that we humans lean as much toward cooperation and generosity as we do toward self-interest under circumstances when we are not coerced by outside forces. Imagine if we created the social and political climate to serve this "Empathic Mutualist Human." I mean, why not?

My lived experience is based in rural landscapes, and I'm embarrassingly aware of my limited perspective on where else gift economies might live side by side with the market. I lecture at college campuses all over the country, so I routinely ask students if and how they participate in gifting networks. I learn about active circles of freecycling, repair cafés, donated mugs in the coffee shop replacing disposables, clothing swaps, the Buy Nothing movement, and campus free stores, where dorm room necessities are passed among generations of students without a penny exchanged. They educate each other on the repercussions of their own consumption and waste. They tell me about the environmental justice issues created by a mega–trash incinerator operating near a community of color. They're fired up by knowing that every item that passes through the free store is diverted from becoming toxic air pollution in a town with a high rate of childhood asthma. These students recognize that their free store makes only a tiny dent in the stream of overconsumption and waste, but it represents a commitment to imagining and practicing an alternative that doesn't pile up injustice along with plastic.

Not surprisingly I suppose, many of these students' examples come from a very different realm—the digital world. They quickly cite access to open-source software and the existence of Wikipedia as manifestations of a gift economy, where knowledge is freely shared on digital platforms in an information commons. Over and over, they name TikTok and YouTube videos where "you can learn anything because someone has made a gift of their time and experience to share with anyone who wants it."

Exploring that digital landscape is as unfamiliar to me as my forests are to them, so I take a field trip to go foraging for videos on gift economies and find them everywhere. I learn about mutual-aid societies, local gift economies, alternative local currencies, money-free work exchanges, cooperative farms, peer-to-peer lending, and more. In these actions, I find both creativity and longing to propel change.

We live in the tension between what is and what is possible. On one hand, we can witness the reciprocity of the economy of nature, showing us how things are supposed to work. And on the other, we see the outcomes of extractive capitalism, breaking every facet of "natural law." I'm sure that I'm not alone in feeling despair in that comparison and my powerlessness to change it. In illuminating these alternatives, people have the courage to say, "Let's create something different, something aligned with our values. We don't have to be complicit."

It is heartening to me that all these grassroots innovations are arising, created to resist the economic systems that are destroying what we love by making new systems founded on protecting what we love. I have a newfound affection for the language we use. It seems apt that we call these "grassroots" movements, emulating the gift economies of plants.

The question of abundance highlights the striking difference between our way of life which has come to dominate the globe and the ancient gift economies which preceded them. There are many examples of functioning gift economies—most in small societies of close relations, where community well-being is recognized as the "unit" of success— where the interest of "we" exceeds that of "I." In this time when economies have grown so large and impersonal that they extinguish rather than nurture community well-being, perhaps we should consider other ways to organize the exchange of goods and services which constitutes an economy.

Most of us are enmeshed in the market economy, which by definition is a monetary system in which the production and distribution of goods is regulated by the "market forces" of supply and demand. Exchanges are voluntary and entrepreneurs are free to pursue profits. The market economy is based on private property and competition in navigating the gap between supply and demand—i.e., scarcity. The greater the gap between supply and demand, the greater the scarcity, and therefore the price to obtain those goods rises and profit increases follow. In a market economy, the meat must be private property, accumulated for the well-being of the hunter or exchanged for currency. The greatest status and success come from possession and profit. Food security is assured by private accumulation.

Gift economies arise from the abundance of gifts from the Earth, which are owned by no one

and therefore shared. Sharing engenders relationships of goodwill and bonds that ensure you will
be invited to the feast when your neighbor is for-
tunate. Security is ensured by nurturing the bonds
of reciprocity. Margaret Atwood writes, "Every
time a gift is given it is enlivened and regenerated
through the new spiritual life it engenders both in
the giver and in the recipient." You can store meat
in your own pantry or in the belly of your brother.
Both have the result of keeping hunger at bay but
with very different consequences for the people
and for the land which provided that sustenance.

M Y ECONOMIST COLLEAGUES remind me that we live in what is known as a "mixed economy," not a purely free market economy. Entrepreneurs are not entirely free to reap profits, they are regulated by the government, where collective agreements are expressed as laws and policies. There is a mixture of private and public goods. Is there room to nurture gift economies within that mix?

Gift economies are everywhere, as we see when we start to pay attention and to name them. Friends invite us to dinner or pass an outgrown stroller to a neighbor with a new baby. A friend of mine makes killer lasagna, which is too much for her, so she always brings some to an elderly neighbor.

The excess in my life tends to be books, because people are always giving them to me. So, when I turn the last page—or sometimes well before—I might give a book to a friend. You do it, too. That simple act is the atom of a gift economy. No money was exchanged, I have no expectation of compensation in any form. That book was kept from the landfill, and my friend and I have a bond and something to talk about; the act of giving opened a channel of reciprocity. It's not so different from what the Serviceberries are doing.

The question that's often asked is how do we take gift economies from individual relationships and scale them up? I have to say that I'm not sure that's the right question. Why does everything have to be expanded? It is the small scale and context that make the flow of gifts meaningful. But if gift economies are to have impact, I'm willing to think about what that might look like on a community scale.

The village down the road from me is too small for a public library. But on a wooden post

outside one of the churches is a brightly painted
box that looks like a house with a glass door. The
two shelves are full of paperback mysteries, chil-
dren's books, and a few how-to manuals. Books
flow in and out to passersby, with no obligation.
Someone's fine woodworking in crafting the box
was a gift to the community, which attracted the
gifts of books, which attracted the gift of readers.
This Little Free Library movement has spread all
over the country to share a love of reading and to
bring books to everyone, in a gift economy. That is
an incremental step beyond sharing a book with a
friend to sharing with your neighborhood.

How does this sharing work in a larger community? Public libraries seem to me a powerful
example of the way that gift economies can coexist
with market economies, at a larger scale. Yes, there
are privately owned bookstores in the city, which
often become meaningful community spaces. I love
bookstores for many reasons but revere both the
idea and the practice of public libraries. To me, they

embody the civic-scale practice of a gift economy and the notion of common property. Libraries are models of gift economies, providing free access not only to books but also music, tools, seeds, and more. We don't each have to own everything. The books at the library belong to everyone, serving the public with free books (and a wider selection than the corner post!). Take the books, enjoy them, bring them back so someone else can enjoy them, with literary abundance for all. And all you need is a library card, which is a kind of agreement to respect and take care of the common good.

As an institution, the library comes close to being a gift economy in the realm of civic life. Although it's not exactly a gift economy since it's supplied by our tax dollars, an involuntary donation to the common good. But it has me wondering whether systems of sharing common public property might be analogues to gift economies.

Libraries, parks, trails, and cultural land-scapes we regard as public goods; they are what we call "common resources"—meant to be shared and cared for by the people who use them. They become possible when we pool our excess dollars in the form of taxes for the common good. We grumble about paying taxes, but in essence this legal obligation is an investment in collective care, in the commons. Other nations provide more than books and green spaces for the common good. These social democracies provide free universal health care, education for all, elder care, family support, and investment in sustainability. The Nordic economies have been termed "cuddly capitalism" in contrast to the "cutthroat capitalism" of the United States. The rate of taxation to support the common good is much higher in these countries than in the United States, but so is the Happiness Index, which in Scandinavia is the highest in the world. When we're paying our taxes, such models help us to imagine how we might encourage elements of gift economies within the matrix of capitalism, which is not going away anytime soon.

The zucchini and flowers model goes beyond individual sharing to organizations, of course. My daughter leads an agricultural extension program for her county, and the Master Gardener members produce lots of surplus from their demonstration plots. Outside their office is a cute little setup where veggies and flowers are distributed for free beneath a sign welcoming neighbors to take what they want. All summer long its bins are full of color and fresh food. After the last frost, when the potatoes are all dug, acorn squashes are gone, and even the hardiest kale has been exhausted, the shelves are bare beneath the sign. It had all been given away and the next day called for storing the stand in the barn for the winter. So they were very surprised to get to work and find it gone! Someone had taken the entire stand. With characteristic generosity, they chalked this up to ambiguous syntax rather than outright theft— after all, the sign did say "Free Farm Stand"!

This is the inherent problem with gift economies—they can't function well when there are cheaters who violate the trust. This little gift economy was derailed by too much taking, by someone breaking the rules of sharing.

Part of the rationale for converting public goods to private commodities resides in the theory of the "Tragedy of the Commons" articulated by Garrett Hardin. His notion is that shared resources—for example, a meadow where all the farmers are allowed to graze their sheep freely—will inevitably be destroyed by competing self-interests. In this theory, the people involved were assumed to behave with the strict self-interest of the "Rational Economic Man." The narrative is that someone will always overgraze or spoil the water source and the collective pasture will become useless to everyone due to the reign of selfishness. Therefore, the story goes, land should be privatized rather than communally held, converting a commonly held source of abundance to individual property in order to safeguard against the Tragedy.

This powerful idea has been used as justification for commodifying what was once understood as a shared gift. But what if it's wrong? What if there's a different story, one that the privateers sought to erase?

The groundbreaking work of economist Elinor Ostrom showed that land can sustain those resources held in common without either state intervention or market economies. Defying the long-held theory, her work showed that collective action, trust, and cooperation can lead to the mutual well-being of land and people without degrading commonly held resources. For this challenge to economic doctrine, Dr. Ostrom was awarded the Nobel Prize in Economic Sciences. This is economics that a botanist can love!

Dr. Ostrom's research grew from her careful observation of systems of land management among communities that colonizing capitalists dismissed as primitive since they did not seem to value or practice accumulation of private property. For much of our human history, before the rise of capital, there were systems in which people viewed land as a common source of abundance, where everyone shared access and met their needs. But it wasn't a free-for-all of unlimited consumption, there were mutual obligations, at scales from individual behavior to international agreements.

For example, in my homelands of the richly forested Great Lakes, Indigenous nations devised what we would call today "resource management plans" for our shared territories. Long before settlement, the Haudenosaunee Confederacy, in lands which are today called New York State and the Anishinaabe Nations of the Great Lakes, made agreements known as the Dish with One Spoon Treaty. This negotiation enshrined the mutual understanding that the lands that provide for both our nations, with hunting grounds and provisions of all sorts, are regarded as One Dish, which Mother Earth has filled with everything that we need to live. It is seen as a gift, and therefore shared in common. There are multiple nations in this accord, but only One Spoon, not a big one for some and a little one for others. It is an agreement for sharing, with a joint responsibility for care.

There are also protocols in many Indigenous cultures for individual taking from the land. These ancient guidelines, referred to as the Honorable Harvest, constrain rampant consumption to ensure that the Dish remains full. I've been told that these ethics themselves are the result of timeless treaties between the Human people and the Deer people, the Bear people, the Fish people, and the Plant people, in which our more-than-human relatives agree to share the gift of their lives to sustain the lives of the humans. In return, the Human people agree to protocols of restraint, respect, and reciprocity. I suspect that these ethics also arose from making mistakes and suffering the ecological consequences. I wonder if we will learn again.

If we think of the Earth as a big warehouse of commodities, as mere objects, we claim a kind of privilege to exploit what we believe that we own.

In that property mindset, how we consume doesn't really matter because it's just stuff and the stuff all belongs to us. There is no moral constraint on consumption. And so, we find ourselves in a time of ecological and spiritual depletion.

But in the worldview of land as gift where the givers are "someones" not "somethings," consumers confront a moral dilemma. We humans must consume, since we are animals to whom the gift of photosynthesis was not given. But our patterns of gross overconsumption have brought us to the brink of disaster. What would it be like to consume with the full awareness that we are the recipients of earthly gifts, which we have not earned? To consume with humility? We are called to harvest honorably, with restraint, respect, reverence, and reciprocity.

The guidelines of the Honorable Harvest are not usually written down, they are reinforced in small acts of daily life. But if I were to list them they would look something like this:

Know the ways of the ones who take care of you, so that you can take care of them.

Introduce yourself. Be accountable as the one who comes asking for a life.

Ask permission before taking. Abide by the answer.

Never take the first one. Never take the last.

Take only what you need.

Take only that which is given.

Never take more than half. Leave some for others.

Harvest in a way that minimizes harm.

Use it respectfully. Never waste what you have taken.

Share.

Give thanks for what you have been given.

Give a gift in reciprocity for what you have taken.

Sustain the ones who sustain you and the Earth will last forever.

I'VE SPENT A LIFETIME asking the plants for their guidance on any number of issues; so I wondered what the Serviceberries had to say about the systems which create and distribute goods and services. What is their economic system? How do they respond to the issues of abundance and scarcity? Has their evolutionary process shaped them to be hoarders or sharers?

The practice of observing the living world and taking inspiration for human ways of living from its model is an essential element of Indigenous science. It embraces the reality that there are intelligences other than our own, from whom we might learn. This ancient mode of knowledge making finds expression today in the emerging science of biomimicry. Janine Benyus and other thinkers are leading a revolution in studying how our economy and our social institutions might be reimagined to align with natural principles rather than against them.

So, let's ask the Saskatoons. These ten-foot-tall trees are the producers in this economy. Using the free raw materials of light, water, and air, they transmute these gifts into leaves and flowers and fruits. They store some energy as sugars in the making of their own bodies, but much of it is shared. Some of the abundance of spring rain and sun manifests in the form of flowers, which offer a feast for insects when it's cold and rainy. The insects return the favor by carrying pollen. Food is rarely in short supply for Saskatoons, but mobility is rare. Movement is a gift of the pollinators, but the energy needed to support buzzing around is scarce. So the trees and the insects create a relationship of exchange that benefits both.

In summer, when the boughs are laden, Service-berries produce an abundance of sugar. Do they hoard that energy for themselves? No, they invite the birds to a feast. Come, my relatives, fill your bellies, say the Serviceberries. Are they not storing their meat in the bellies of their brothers and sisters—the Jays, the Thrashers, and the Robins?

Isn't this an economy? A system of distribution of goods and services that meets the needs of the community. The currency of this economic system is energy, which flows through it, and materials, which cycle among the producers and the consumers. It is a system for redistribution of wealth, an exchange of goods and services. Each member has an abundance of something, which they offer to others. The abundance of berries goes to the birds—for what use are berries to the tree other than as a way to make relationships with birds?

Eating too many berries has the same effect on birds as it does on people. Fuchsia splats decorate the fence posts. This of course is the whole point of berries—to make themselves so irresistible and plentiful that birds will come and feast, as we pickers are doing this evening, and then distribute the seeds far and wide. Feasting has another benefit. Passage through a bird gut scarifies the seeds to stimulate germination. The birds provide services to the Serviceberries, who provide for them in return. The relationships created by the gift weave myriad relations between insects and microbes and root systems. The gift is multiplied with every giving, until it returns so rich and sweet that it burbles forth as the birdsong that wakes me in the morning. If the abundance had been hoarded, if Juneberries acted solely for their own benefit, the forest would be diminished.

This makes me think about the person who stole the Free Farm Stand. We had laughed it off as a mistake, of little consequence. But the mindset that claims permission to convert a free gift to private property, robbing the community for individual gain, is of the highest consequence. Our petty thief deserves a name, so let's call him Darren, after the CEO of ExxonMobil. We often lay the blame for the outcomes of cutthroat capitalism on the "System." There's merit in that, given the complex layered interactions, but no excuse. Let's remember that the "System" is led by individuals, by a relatively small number of people, who have names, with more money than God and certainly less compassion. They sit in boardrooms deciding to exploit fossil fuels for short-term gain while the world burns. They know the science, they know the consequences, but they proceed with ecocidal business as usual and do it anyway. Their behavior feels to me like the same kind of arrogant entitlement as Darren the Farm Stand thief or Darren the Planet Wrecker. They're all thieves, stealing our future, while we pass around the zucchini.

I'm not minimizing the significance of our sweet gift economies. On the contrary, they are emblems of the rich land-based abundance that is possible, stitched together with what was once called being a good neighbor. There are so many more of us than of the Darrens, and an asymmetry in what they recognize as power.

I lament my own immersion in an economy that grinds what is beautiful and unique into dollars, converts gifts to commodities in a currency that enables us to purchase things we don't really need while destroying what we do.

The Serviceberries show us another model, one based upon reciprocity rather than accumulation, where wealth and security come from the quality of our relationships, not from the illusion of self-sufficiency. Without gift relationships with bees and birds, the Serviceberries would disappear from the planet. Even if they hoarded abundance, perching atop the wealth ladder, they would not save themselves from the fate of extinction if their partners did not share in that abundance. Hoarding won't save us either. It won't even save Darren. All flourishing is mutual.

As I watch the Robins and Cedar Waxwings fill their bellies, I see a gift economy in which abundance is stored "in the belly of my brother." Supporting a thriving bird community is essential to the well-being of the Serviceberry and everyone else up and down the food chain. That seems especially important to an immobile, long-lived being like a tree, who can't run away from ruptured relationships. Thriving is possible only if you have nurtured strong bonds with your community.

I'm a plant ecologist, so I wonder whether an economist like Valerie would see a gift economy in Serviceberry's distribution of goods and services. I want to know if natural systems could be understood as analogues to economic systems. Could we engage in a kind of biomimicry to design systems of exchange which benefit human people and more-than-human people at the same time?

"Yes!" Valerie says, as if she'd been waiting a long time to be asked this question. "Natural systems can surely be understood as analogues to economic systems." Here again, is the premise of biomimicry.

Imagining human economies that are modeled after ecological systems is the realm of ecological economists like Valerie. Ecological economists ask how we might build economic systems that meet citizens' needs while aligning with ecological principles that allow long-term sustainability for people and for the planet. Valerie says that "ecological economics emerged after observing [how the] neoclassical economic approach fails to provide for everyone and does not adequately consider the ecosystems that are our life support. We've created a system such that we self-identify as consumers first before understanding ourselves as ecosystem citizens. In ecological economics, the focus is on creating an economy that provides for a just and sustainable future in which both human life and nonhuman life can flourish."

What might Serviceberry teach us here? She replies, "Serviceberry, or Shadbush as I learned it, provides a model of interdependence and coevolution that is the heart of ecological economics. Serviceberry teaches us another way to understand relationships and exchange. A Serviceberry economy as our model prompts the opportunity for articulation of the value of gratitude and reciprocity as essential foundations for an economy." Reciprocity—not scarcity.

As a participant in a traditional culture of gratitude, now with a bucket full of berries in my hand, I've never quite understood something about human economics, and that is the primacy of scarcity as an organizing principle.

As a person schooled by plants, my fingers stained with berry juice, I'm not willing to give scarcity such a prominent role. Gift economies arise from an understanding of earthly abundance and the gratitude it generates. A perception of abundance, based on the notion that there is enough if we share it, underlies economies of mutual support.

Valerie points out that even ecologists are reevaluating the assumption that intense competition is the primary force regulating evolutionary success. Evolutionary biologist David Sloan

Wilson has found that competition makes sense only when we consider the unit of evolution to be the individual. When the focus shifts to the level of a group, cooperation is a better model, not only for surviving but for thriving. In a recent interview, author Richard Powers comments, "There is symbiosis at every single level of living things, and you cannot compete in a zero-sum game with creatures upon whom your existence depends." Serviceberries discovered this long ago, and we humans need to catch up. And yet, we continue to operate from the foundation of competition.

There is no question that all living beings experience some level of scarcity at various points, and therefore that competition for limited resources, like light or water or soil nitrogen, will occur. But since competition reduces the carrying capacity for all concerned, natural selection favors those who can avoid competition. Oftentimes this avoidance is achieved by shifting one's needs away from whatever is in short supply, as though evolution were suggesting "If there's not enough of what you want, then want something else." This specialization to avoid scarcity has led to a dazzling array of biodiversity, each species avoiding competition by being different. Diversity in ways of being is an antidote.

Some evolutionary biologists might reject this notion, framing the lifeways of Serviceberry as maximizing self-interest through natural selection, which is the same sort of argument made by market economists. Competition between individuals for success was long seen as the driving force in ecology and in economics. Science, politics, and economics were entangled in adopting metaphors from the natural world that reflected as much about social attitudes as about ecological reality. But that approach has been increasingly questioned, and scientific evidence is mounting that mutualism and cooperation also play a major role in evolution and enhance ecological well-being, especially in changing environments. Mutualism or reciprocal exchanges create abundance for both partners, by sharing.

The Serviceberries are networked not only aboveground with partners for pollination and dispersal but belowground with webs of mycorrhizal fungi and other microbial communities that are exchanging resources. Perhaps inculcated with the Tragedy of the Commons perspective, we used to assume that these fungi were "stealing" nutrients from the trees, but the closer we look, it seems as if the nutrients might be freely given in a network of reciprocity.

What if scarcity is just a cultural construct, a fiction that fences us off from a better way of life? When I examine Serviceberry economics, I don't see scarcity, I see abundance shared: photosynthate is usually not in short supply, since sun and air are perpetually renewable resources. Of course, sometimes there's not enough rain, and then the scarcity ripples through the web of relationships, for sure. That is real scarcity: when the rains don't come. It is a physical limitation with repercussions and loss that are shared, just as abundance is shared. That kind of scarcity, produced by natural fluctuation is not what troubles me.

It is manufactured scarcity that I cannot accept. In order for capitalist market economies to function, there must be scarcity, and the system is designed to create scarcity where it does not actually exist. Because I had not thought much about economics since my introduction to it in high school decades ago, I realize that I had just been accepting the principle of scarcity as if it were a natural fact.

I try to lay out the understanding for myself, to think like an economist, not an ecologist. In order for money to be made, there must be commodities to be bought and sold. The scarcer those commodities, the higher their cost and thus the greater the revenue. So, I guess I do understand this: market economics demands that abundant, freely available, earthly gifts be converted to commodities and made scarce by privatization and high prices. This seems crazy, so let me test my understanding with the example of pure, beautiful water, a gift from the skies. It was previously unthinkable that one would pay for a drink of water; but as careless economic expansion pollutes fresh water, we now incentivize privatization of springs and aquifers. Sweet water, a free gift of the Earth, is pirated by faceless corporations who encase it in plastic containers to sell. And now many can't afford what was previously free, and we incentivize wrecking public waters to create demand for the privatized. What induces people to buy bottled water from a corporation more convincingly than contaminated water flowing from the faucet?

In contrast, in Indigenous societies all over the world, where remnants of gift economies endure, water is sacred and people have a moral responsibility to care for it, to keep it flowing. It is a gift, to be shared by all, and the notion of owning water is an ecological and ethical travesty. Lewis Hyde writes, "To assign a market value to a gift, destroys the gift."

Continued fealty to economies based on competition for manufactured scarcity, rather than cooperation around natural abundance, is now causing us to face the danger of producing real scarcity, evident in growing shortages of food and clean water, breathable air, and fertile soil. Climate change is a product of this extractive economy and is forcing us to confront the inevitable outcome of our consumptive lifestyle: genuine scarcity, for which the market has no remedy.

The Indigenous philosophy of the gift economy, based in our responsibility to pass on those gifts, has no tolerance for creating artificial scarcity through hoarding. In fact, the "monster" in Potawatomi culture is Windigo, who suffers from the illness of taking too much and sharing too little. It is a cannibal, whose hunger is never sated, eating through the world. Windigo thinking jeopardizes the survival of the community by incentivizing individual accumulation far beyond the satisfaction of "enoughness." Contemporary Windigos who cannibalize life for accumulation of money need their own name. Perhaps "Darren" would fit.

THE THREAT OF REAL SCARCITY on the horizon is brought to us by unbridled capitalism. Extraction and consumption outstrip the capacity of the Earth to replenish what we have taken. An economy based on the impossibility of ever expanding growth leads us into nightmare scenarios. I cringe when I hear economic reports celebrating the accelerating pace of economic growth, as if that were a good thing. It might be good for the Darrens, for the short term, but it is a dead end for others—it is an engine of extinction.

At one point while writing—as I was struggling to imagine how the model of Serviceberries and ancient gift economies could help us imagine our way out of the mutually assured destruction of cutthroat capitalism—I needed a break from the Windigo shadows that were creeping toward me. Thankfully, I was interrupted by a text from my neighbor Paulie. As though she were reading my troubled mind from across the valley, Paulie invited me to come pick berries at her farm. Serviceberries. For free. The tingles of synchronicity propelled me from my desk to the orchard.

Paulie and Ed Drexler run Springside Farm. I can see the rows of Christmas trees, the corn maze, and the pumpkin patch from here. Paulie planted this orchard with commodity in mind, part of her revenue stream as a small local farmer; an innovative crop destined for "pick your own" fees, which can be lucrative. But instead, that day she invited her neighbors to come and pick for free. Her labor and expenses are not free: tilling, irrigation, and marketing cost real money. The trees cost money, as does the gas when Ed mows between rows—and the Serviceberries won't be paying their own way.

She is losing the return on her investment by inviting us to come fill our buckets with this surfeit of sweetness. She is not obeying the rules of the capitalist market economy; she is not behaving in a way that will maximize her profit. How un-American.

In one fell swoop her berries rolled from the commodity column on a spreadsheet into the beribboned box called "gift." The berries hadn't changed a bit: they were still juicy and bursting with antioxidants. The farm hadn't changed either. It's a small family operation, diversified with an array of crops that generate revenue all year—from early spring lambs to Christmas trees. The only thing that changed was whether the folks who came to pick berries were asked to put pieces of green paper into the coffee can inside the barn door.

I asked her why she did it, especially in these pandemic days, when every small business is struggling to make ends meet. "Well," she said, "they're so abundant. There's more than enough to share, and people could use a little goodness in their lives right now." People came to pick in the cool of the early evening, socially distanced at opposite ends of the rows, isolated yet somehow connected by the rhythm of fingers moving from bush to bucket—and mouth. "Everyone's so sad now," she said, "but in the berry patch all I hear are happy voices. It feels good to give that little bit of delight."

But it's also education, she says. Most people don't know Juneberries yet and giving them away is an invitation to try them. Now used for making pies and jam and cramming your mouth full, they are celebrated as a gift from the land but are little known as a product in the market economy. Paulie says her goal is simply to get Juneberries into someone's mouth for the first time; the berries will do the rest.

Paulie has a reputation to uphold for being no-nonsense in her approach to life, so she qualifies her explanation: "It's not really altruism," she insists. "An investment in community always comes back to you in some way. Maybe people who come for Serviceberries will come back for Sunflowers and then for the Blueberries. Sure, it's a gift, but it's also good marketing. The gift builds relationships, and that's always a good thing. That's what we really produce here—relationship, with each other and with the farm." The currency of relationship can manifest itself as money down the road, because Paulie and Ed do have to pay the bills. Free berries might translate to better pumpkin sales, because people will want to come back to a place they have a relationship with. "People feel like they got something more than they paid for," she explained. "They learned about a new food, or watched the kids climb on hay bales." Good feelings are the real value added. Even when something is paid for as a commodity, the gift of relationship is still attached to it.

The ongoing reciprocity in gifting stretches beyond the next customer, though, into a whole web of relations that are not transactional. Paulie and

Ed are banking goodwill, so-called social capital. "Being known as a citizen is always of value," she says. If someone leaves a gate open and her sheep end up in my garden, there's a cushion of goodwill in place so that the munched dahlias may be forgiven. "The way I see it," she says, "always value people over things. There's that old line that farmers like to spout, 'Without farmers, you'd be naked, hungry, and sober.' But it goes both ways: without good neighbors, you'd also be alone, and that's worse."

And that customer who comes to value the smell of ripe berries and the view of lambs on pasture and the memory of their kids climbing on hay bales—they just might vote for the farmland preservation bond in the next election. That's a fine return on investment from a free bucket of berries.

I cherish the notion of the gift economy, that we might back away from the grinding system, which reduces everything to a commodity and leaves most of us bereft of what we really want: a sense of belonging and relationship and purpose and beauty, which can never be commoditized. I want to be part of a system in which wealth means having enough to share, and where the gratification of meeting your family needs is not poisoned by destroying that possibility for someone else. I want to live in a society where the currency of exchange is gratitude and the infinitely renewable resource of kindness, which multiplies every time it is shared rather than depreciating with use.

Anthropologists who study gift economies note that they function well in small, tightly knit communities. You might rightly observe that we no longer live in small, close-knit societies, where generosity and mutual esteem structure our relations. But we could. It is within our power to create such webs of interdependence, quite outside the market economy. Maybe that is how we extract ourselves from a cannibal economy. Intentional communities of mutual self-reliance and reciprocity are the wave of the future, and their currency is sharing. The move toward a local food economy is not just about freshness and food miles and carbon footprints and soil organic matter. It's about all of those things, but it's also about the deep human desire for connection, for honor, to be in reciprocity with the gifts that are given you.

The real human needs that such arrangements address are exactly what we long for yet cannot ever purchase: being valued for your own unique gifts, earning the regard of your neighbors for the quality of your character, not the quantity of your possessions; what you give, not what you have.

I don't think market capitalism is going to vanish; the faceless institutions that benefit from it are too entrenched. The thieves are very powerful. But I don't think it's pie in the sky to imagine that we can create incentives to nurture a gift economy that runs right alongside the market economy. After all, what we crave is not trickle-down, faceless profits but reciprocal, face-to-face relationships, which are naturally abundant but made scarce by the anonymity of large-scale economics. We have the power to change that, to develop the local, reciprocal economies that serve community rather than undermine it.

In *Sacred Economics*, Charles Eisenstein reflects on the economy of ecosystems: "In nature, headlong growth and all-out competition are features of immature ecosystems, followed by complex interdependency, symbiosis, cooperation, and the cycling of resources. The next stage of human economy will parallel what we are beginning to understand about nature. It will call forth the gifts of each of us; it will emphasize cooperation over competition; it will encourage circulation over hoarding; and it will be cyclical, not linear. Money may not disappear anytime soon, but it will serve a diminished role even as it takes on more of the properties of the gift. The economy will shrink, and our lives will grow."

You know, the saga of the Little Free Farm Stand offers another glimpse of what is possible.

Yes, it was taken, disrupting a nascent gift economy by "privatizing" a gift, also known as stealing. But, the following spring, a local Eagle Scout volunteered to build a new one. In fact, he plans to make several and place them throughout the community for freely sharing vegetables. There will be Little Free Farm Stands beside the Little Free Libraries. He is disrupting the model of a market economy and earning honor for supporting an alternative. No longer will gardeners have to sneak surplus zucchini into strangers' mailboxes. The surplus will be stored in the belly of a neighbor.

I've been learning more about ecological economics, valuing of ecosystem services, biomimicry, proposals for climate justice, for climate finance, for the Green New Deal, for energy currencies and B Corps—not that I want to. The language is as obscure to me as my botanical terminology is to economists. The more I learn about the visionary proponents of regenerative economies, the more immensely grateful I am to the brilliant folks who are working toward creating a different system, one that acts on behalf of a livable future.

Kate Raworth's well-known model of "Doughnut Economics" comes to mind. She challenges the flawed assumptions of contemporary economics and proposes instead an economy that is bounded above by ecological limits and rests on a foundation of social justice. She writes that thriving depends on more than meeting basic physical needs, and includes goods like a sense of community, mutual support, and equality. Wealth is much more than what GDP measures, and the market is not the only source of economic value. She urges policymakers to recognize the values of common lands, green space, biodiversity. Her models incorporate the "productivity" of unpaid labor like family care and volunteering, growing a garden—those elements of thriving that never appear in a spreadsheet, but are essential to our wellbeing.

Likewise, Katherine Collins has become an outspoken architect of investment strategies that help propel us toward a circular economy. She took the remarkable step in her business career of going to divinity school, so that her vocabulary of values was as strong as the language of finance. That made me want to listen. It seems to me that these thinkers have been greatly influenced by what the Service-berry already knows and has been showing us from the start. And the Maples and the Cattails and the Dandelions. But we replaced their wisdom with equations of our own making. I was gratified to learn that Dr. Raworth now incorporates the Honorable Harvest into her economics class at Oxford University. Change is coming.

In these urgent times of impending climate catastrophe, we need to move quickly to the decarbonized economy that is essential to life as we know it. A fellow tribal member captured my thoughts when she wrote, "If the economy requires people to consume more resources than the Earth can replenish, just to keep the whole thing from collapsing, isn't it time for a new economy?"

But how does a new system come to replace an entrenched one?

As a botanist, I know that there is guidance from the world of fields and forests. Plant communities are changing and replacing one another all of the time, in a dynamic mosaic we know as ecological succession. Far from the stereotype of the "forest primeval," plant communities are constantly in flux. From a bird's-eye view, the "unbroken forest" is in fact a patchwork of stands of different ages and experience. Fires, landslides, floods, windstorms, outbreaks of insects, disease, and disasters of human origin disrupt the green blanket in unpredictable ways—and yet with a somewhat predictable response. Oftentimes a major disturbance that clears the former forest creates a gap, with full sun, disturbed soil, and plenty of resources, since the previous inhabitants are now gone. Such places are colonized by fast-growing species in high density, trying to take advantage of the transitory conditions. These pioneer species are

opportunists, with traits that consume resources, crowd out others, and reproduce like crazy. It's all "me, me, me," investing only in their own exponential growth with no regard for the future, their relatives, or longevity. Sound familiar? It's a field of fast-growing weeds, or a stand of aspens. It's as if Euromericans, in the age of colonization and displacement of "old-growth cultures" are behaving like colonizing plants after a massive disturbance, dominating the landscape. But those colonizing plants find they cannot continue this rate of growth and resource extraction. They start to run out of resources, disease may attack the overdense populations, and competition begins to limit their growth. In fact, their behavior facilitates their own replacement. Their rampant growth captures nutrients and builds the more stable conditions in which their followers can flourish. Incrementally, they start to be replaced.

The ones who come next are different, growing more slowly in a resource-limited world. Stressful

conditions incentivize nurturing relations of cooperation alongside competition. The extractive
practices of the colonists must be replaced with
reciprocity and replenishment if anyone is to sur-
vive. Investing in persistence, the new inhabitants
are in it for the long haul. These communities have
been called "mature" and sustainable, in contrast to
the adolescent behavior of their predecessors. This
transition from exploitation to reciprocity, from
the individual good to the common good has been
seen as a parallel to the transition that colonizing
human societies must undergo, from hoarding to
circulation, from independent to interdependent,
from wounding to healing, if we are to thrive into
the future.

How do systems change? How can we move
toward the just communities we need and want?
The natural process of ecological replacement
highlights two mechanisms at work in replacing
a complex system that dominates the landscape and
seems too big to change. Succession relies in part on

incremental change, the slow, steady replacement of that which does not serve ecological flourishing with new communities. But it also relies on disturbance, on disruption of the status quo in order to let new species emerge and flower. Some massive disturbances are destructive, and recovery from them may not be possible. Other disturbances, of the right scale and type, create renewal and diversity. Indigenous land stewardship relied on humans using carefully calibrated kinds of disturbance to create a living mosaic in different stages of recovery. Disruptions create gaps, openings and edges between the new and the dominant. I want to see emerging gift economies nurtured in the gaps carved out of the overbearing market economy.

Both of these tools—incremental change and creative disruption—are available to us as agents of cultural transformation. I hope we will use them both. In these urgent times, we need to become the storm that topples the senescent, destructive economies so the new can emerge. The gap edges, or ecotones, where two ecosystems, the new and the old, meet, are among the most diverse and productive of ecosystems, full of berries and birds. There are species that live in neither the new nor the old, but at the edges. This is the home of Cedar Waxwings and the realm of Serviceberries.

The Darrens' economy of extractive capitalism, of abusing the gifts of Mother Earth, is a crime against Nature. I believe that theft is punishable by law, and we need to elect leaders who believe in the rule of law. The fossil fuel economy is propelling mass extinction in acidified oceans and disappearing forests, deadly heat waves and untold human suffering. How far down the species-at-risk list are Cedar Waxwings and Serviceberries? I fear for the well-being of my sweet green valley and the livelihoods of small farmers.

Already the land is too quiet. What if our metrics for well-being included birdsong, the crescendo of Crickets on a summer evening, and neighbors calling to each other across the road?

I see the potential for a mosaic of economies emerging in the example of my neighbors. Yes, they have to pay the bills and are part of the market economy, but they participate in a gift economy at the same time. With every product sold they add something that cannot be commodified, and that is therefore even more valuable. People come to them for a sense of connection to the land, a laugh with the farmer as a fellow human who cherishes the crisp fall air—not for the commodity of a pumpkin, which, after all, they could buy anywhere. Gift economies are more fun, more satisfying, and as nourishing as the Juneberry pancakes at Springside Farm. And sneaking zucchini into a parked car. I've long believed that the ones who have more joy win. Maybe this levels the power differential between us and Darren. We have joy and justice on our side. And berries.

Ed and Paulie cultivate other berries here, too. They are famous for blueberry pancake breakfasts in the height of summer that bring neighbors to the farm. As the berries plunked into my bucket, I reflected on my long-held belief that berry picking is the first step in a lifelong companionship with the living world. I've seen it happen. Students on my walks who hold themselves back, flash their skepticism of gift thinking with barely concealed eye rolls. Too cool for school, there's no way they're going to put a wild wintergreen leaf in their mouths. But when we get to the raspberry bushes, I know they'll have to drop their guard. The simple act of encountering a wild berry, dangling there just waiting for their fingers and their mouth, loosens something in them, to the evidence of the gift. I've begun to think that berry-picking is the medicine we need to create a legion of land protectors.

I have some family members who have taken this a step farther. They live in an urban neighborhood where there are plenty of curmudgeons growling at kids to stay off their lawns. So they converted their once tidy little yard to a garden of berries and patches of flowers and put up a welcome sign for the kids in the neighborhood to come on in and find a handful of berries and pick a bouquet to take home. They converted their "private" yard to a common space. The currency of this gift economy is relationship and a neighborhood where people know each other's names, even the curmudgeons. The Tragedy of the Commons became the Abundance of Community. This is a gift economy in reach of everyone. It's subversive. And delicious.

Regenerative economies that reciprocate the gift are the only path forward. To replenish the possibility of mutual flourishing, for birds and berries and people, we need an economy that shares the gifts of the Earth, following the lead of our oldest teachers, the plants. They invite us all into the circle to give our human gifts in return for all we are given. How will we answer?

AN INVITATION TO BE A PART
OF THE GIFT ECONOMY

The author's advance payments from this book about the gift economy of the natural world will be donated as a reciprocal gift, back to the land, for land protection, restoration, and justice in support of healing land and people.

In the spirit of the reciprocal gift economy, you might consider how you can reciprocate the gifts of the Earth in your own way. Whatever your currency of reciprocity—be it money, time, energy, political action, art, science, education, planting, community action, restoration, acts of care, large and small—all are needed in these urgent times. You're invited to become a member of the gift economy on behalf of people and planet.

ACKNOWLEDGMENTS

There's nothing in this world that we do alone. I offer my gratitude to the many folks whose thoughts and actions have made this little book possible. Ed and Paulie Drexler of Springside Farm provided the setting of their wonderful agritourism farm, which invites people to get to know the gifts of the land and the gift of good neighbors. I appreciate the patience of my son-in-law Dave for talking with me about economics, of which my understanding is rudimentary at best. My friend Dr. Valerie Luzadis was so helpful with her always-perceptive conversations. My daughter Larkin shared the story of the Little Free Farm Stand and hopes for its restoration. I'm grateful to Miki Kashtan and Madi Loustalot for introducing me to the language of the maternal gift economy—and to my mother and daughters for living it.

This essay was originally published in *Emergence Magazine*. Thank you for your permission to include and expand upon it here. I'm so grateful for the support of the MacArthur Foundation for helping to make the time and space to create it. It has been a pleasure to work with my editor Chris Richards, and gratitude for inviting this slim volume into being. I appreciate the fine artwork of the illustrator John Burgoyne. Much gratitude for the care and guidance of Christie Hinrichs of Authors Unbound and to Sarah Levitt of Aevitas Creative and Chloe Currens of Allen Lane.

Every single day I am grateful for the love, support, and inspiration of my family and friends, who make this life possible. Special gratitude goes out to the birds and the berries: G'chi megwech.

ABOUT THE AUTHOR
AND ILLUSTRATOR

Robin Wall Kimmerer (*author*) is a mother, scientist, professor, and enrolled member of the Citizen Potawatomi Nation. She is the author of the #1 *New York Times* bestseller *Braiding Sweetgrass: Indigenous Wisdom, Scientific Knowledge, and the Teachings of Plants* as well as *Gathering Moss: A Natural and Cultural History of Mosses*. Kimmerer is a 2022 MacArthur Fellow. She lives in Syracuse, New York, where she is a SUNY Distinguished Teaching Professor of Environmental Biology and the founder of the Center for Native Peoples and the Environment.

John Burgoyne (*illustrator*) is a member of the New York Society of Illustrators and an alumnus of Massachusetts College of Art. John has won over 100 awards in the United States and Europe, including from the Society of Illustrators, Communication Arts, Hatch Awards, Graphis, Print, One Show, New York Art Directors Club, and Clio. His work can be found at JohnTBurgoyneIllustration.com.